Maurício Dominguez Nasser

Propagation by forking of the acerola tree

Maurício Dominguez Nasser

Propagation by forking of the acerola tree

cv. Okinawa on different rootstocks

ScienciaScripts

Cover image: www.ingimage.com

This book is a translation from the original published under ISBN 978-613-9-65770-4.

Publisher:
Sciencia Scripts
is a trademark of
Dodo Books Indian Ocean Ltd. and OmniScriptum S.R.L publishing group

120 High Road, East Finchley, London, N2 9ED, United Kingdom
Str. Armeneasca 28/1, office 1, Chisinau MD-2012, Republic of Moldova, Europe
Printed at: see last page
ISBN: 978-620-7-79966-4

SUMMARY

"To the people who work in agriculture, and that in some way this work can help improve their lives, I dedicate it."

"Sustainability is more than understanding knowledge, it's practicing it."

DEDICATE

To my parents José Nasser Sobrinho and Luisa Dominguez Nasser for the eternal love, support, patience, affection and guidance they have given me throughout my life.

And also to my siblings Luciana Dominguez Nasser and Paulo Dominguez Nasser for our love, as well as the joys and sorrows we have experienced together, and may our union be infinite.

To my wife Flâvia Aparecida de Carvalho Mariano Nasser for her love and complicity.

ACKNOWLEDGMENTS

First of all, to God, creator of heaven and earth, for the strength, light, wisdom and courage that have enabled me to complete this stage of my life.

To my parents and siblings for their love, understanding and total encouragement throughout my work.

To my dear wife and girlfriend Flâvia Aparecida de Carvalho Mariano Nasser for her teachings, support, affection, dedication and patience.

To all my family members who have supported me and who in some way are always thinking positively of me.

To the Alta Paulista Regional Center of the São Paulo Agency for Agribusiness Technology (APTA), which gave me the opportunity to study for a Master's Degree in Agronomy at the Ilha Solteira School of Engineering (Universidade Estadual Paulista "Jûlio de Mesquita Filho"), Ilha Solteira Campus.

To the Postgraduate Program in Agronomy at the Faculty of Engineering of UNESP - Ilha Solteira Campus for giving me the opportunity to study for a Master's degree.

To Professor Dr. Luiz de Souza Corrêa for accepting me as his advisor; for his guidance and wise words at all times during this challenge called the Master's Degree in Agronomy.

Professor Dr. Aparecida Conceiçao Boliani for her simple but true teachings in professional and personal life.

To Professor Dr. Pedro César dos Santos for the statistically significant conversations.

To the Agricultural Engineer and Scientific Researcher Dr. José Carlos Cavichioli, who encouraged and influenced me to research Tropical Fruticulture, and for the attention he gave me before and during this academic journey.

To Takashi Yokoyama, an agronomist, for his support and ideas on how to apply them to the aceroleira crop, and for passing on some of his experience as an extension worker and farmer.

To Marcelo Rosa Melo, an agricultural engineer, for his support and for setting up the Active Germplasm Bank of aceroleira located in Adamantina, the headquarters of APTA Regional Alta Paulista, which served as the base for the experiment.

To my postgraduate colleagues Maximiliano, Vandervilson, Haroldo, Antònio, Maria Cecilia, Veridiana, Èrica, Gustavo, Marcus, for their friendship and partnership in the work carried out during the course.

To all the teachers, students and staff who have been part of this journey throughout the course, and who, for whatever reason or way of acting, have shown me how to be a better professional and human being,

To all my friends from short and long ago, my work colleagues at the Alta Paulista Regional Pole / APTA, the Northeast Paulista Regional Pole / APTA and the Agronomy Postgraduate course for the discussions, laughter, complicity, conviviality and, above all, friendship.

EPILOGUE

"People don't need to be geniuses or extraordinarily talented to do their jobs. The requirements are twofold: initiative and diligence, qualities that abound among humans when they are not discouraged or suppressed."

José Eli da Veiga, 2005 (excerpt from the book: Desenvolvimento Sustentável: o desafio do Século XXI).

SUMMARY

The importance of studying acerola is due to the high content of ascorbic acid (vitamin C) in the fruit, with values of up to 4000 mg per 100 g of pulp. The area occupied by the acerola tree (*Malpighia emarginata Sessé e Mocino ex DC.*) in the state of Sao Paulo covers 597 ha and 65% of production is located in the Alta Paulista region. In this region, producers use the Olivier cultivar, and propagation is carried out by cuttings, due to the low germination of seeds and the plants' lack of uniformity when propagated by seeds. In order to encourage the cultivation of other varieties of acerola, such as Okinawa, which stands out in other producing regions of Brazil, and in the search for an alternative rootstock for nematodes, the aim of this study was to evaluate the setting and initial development of seedlings of the Okinawa acerola cv. propagated by full cleft grafting on different rootstocks. The experiment was conducted from January 10 to December 12, 2011 in the experimental nursery of the APTA Alta Paulista Regional Center in Adamantina-SP. The treatments consisted of 7 rootstocks propagated by seed, of the following cultivars: Okinawa, Olivier, Apodi (BRS 235), Cereja (BRS 236), Roxinha (BRS 237), Frutacor (BRS 238) and Waldy CATI 30. Only forks of the Okinawa cultivar were used for grafting. The experimental design was entirely randomized with 7 treatments (rootstocks) and 3 replications. At 40, 55 and 75 days after grafting, the following variables were evaluated: percentage of setting through the emission of shoots above the grafted region; height of the seedling, diameter of the rootstock and scion stems, number of leaves and shoots emitted from the scion. At 75 days after grafting, the dry matter mass of the aerial part, roots and total was evaluated. Based on the results obtained, it can be concluded that, due to the high percentage of grafting success on the different rootstocks, this technique proved to be viable for the crop. There was no difference between the different scion/rootstock combinations in terms of the initial development of the seedlings. The total time from sowing to the final stage of production of the grafted acerola seedlings was 11 months, and at 75 days after grafting they were in a condition to be planted in the field.

KeywordsMalpighiaemarginata D. C.Grafting.setting index.

1. INTRODUCTION

Worldwide, fruit production is growing and is made up of various species, most of which are grown in temperate climates. China, India, Brazil and the United States contribute 46.4% of the total produced (ANDRADE, 2012). In third place in world fruit production, Brazil produced more than 44 million tons (FAO, 2011).

Naumov (2009) reports that tropical fruits are also in high demand, as they have high nutritional value and a special flavor. The author states that Brazil is considered one of the largest producers of tropical fruit in the world. In this sense, the importance of studying acerola was initiated by Asenjo and Freire de Guzman (1946), when they detected values of up to 4000 mg of ascorbic acid per 100 g of pulp.

The acerola tree is grown mainly in the following countries: Brazil, Puerto Rico, Cuba and the United States; and to a lesser extent in Venezuela, Colombia, some Caribbean islands and Asian countries (CARDOSO; LOPES; ALMEIDA, 2003). In Brazil, 24,451 tons of acerola were produced on 3,494 hectares, with the Northeast region and the state of São Paulo standing out (IBGE, 2006).

In the state of São Paulo alone, the crop occupies 597 ha, 260 ha of which are located in the Alta Paulista region, especially in the municipality of Junqueirópolis (CATI, 2008).

Characterized as a perennial crop, the main varieties grown are Okinawa, Sertaneja and Flor Branca in the states of Minas Gerais, Sergipe, Bahia and Pernambuco, mainly in the Sâo Francisco River Valley. In São Paulo, in the Junqueirópolis region, the Olivier cultivar was selected for planting by a producer and is considered suitable for pulp and table production (ADRIANO; LEONEL; EVANGELISTA, 2011; RITZINGER; RITZINGER, 2011).

On a commercial scale, the most widely used propagation method is cuttings, as it has provided greater speed in production and guarantees the maintenance of the genetic characteristics of the parent plant (KLUGE; REZENDE, 2003). However, in acerola growing, when selected clones are available, propagation by cuttings or grafting is recommended (KAWATI, 1995).

In the acerola crop, in general, the use of grafting with seed-formed rootstocks is a vegetative propagation that provides a more vigorous root system (ALVES et al. 2009).

Considering that grafting studies on acerola are scarce, it is believed that the rootstock can influence the vigor, longevity, nutrient deficiency, health, quality and ripeness of the fruit, just as it happens with other fruit trees (OLIVEIRA, et al. 2003b; SASSO; CITADIN; DANNER, 2010; SIMAO, 1998; PAIVA, 2003a).

The aim of this study was to evaluate the setting and initial development of Okinawa acerole seedlings grafted onto different rootstocks.

2. LITERATURE REVIEW

2. 1 Origin and geographical distribution of the acerola crop

The acerola, *Malpighia emarginata Sessé e Mocino ex DC.,* belongs to the Malpighiaceae family and originated in northern South America, the Caribbean Islands and Central America (COUTO, 1997; JANICK; PAULL, 2008). Marino Netto (1986), considered the aceroleira to come from the Antilles, northern South America and Central America, but did not find clarification on its center of origin.

Couceiro (1985) reported that the dispersal of the acerola occurred long before the discovery of America through the natives of the islands of Central America who fed on this fruit and took it as food on their way from one country to another, spreading it to the various islands.

In Brazil, it has been grown since the 1980s, mainly in the Northeast. In 2003, approximately 10,000 tons of acerola fruit were processed by the Bahia industries Utiara, Brasfrut, Companhia de Citricos do Brasil - Cajuba (CCB-Cajuba), and other smaller companies (RITZINGER; RITZINGER, 2011). Oliveira, Soares Filho and Cunha (1998) estimated Brazilian production at around 150,000 tons of fruit, especially in the Northeast.

Of the producing states, Pernambuco stands out as the largest producer of acerola with 7700 tons, and together with the states of Paraiba, Sao Paulo, Bahia and Cearà produced more than 75% of the acerola grown in the country (IBGE, 2006).

The Agricultural Association of Junqueirópolis-SP, 2010, which has been producing acerola since 1991 (FREITAS; VILPOUX, 2007), and sells around 4,000 tons of acerola every year, found through its representatives that demand for the product is constant and that the trend for the coming years is for an increase in terms of the export market (verbal information).[1] On the 175 hectares cultivated with acerola, 60% of the fruit is exported to Japan and Europe as frozen pulp, but it also serves the domestic market, including school lunch programs. The crop employs 500 people during the harvest months from October to April next year (O ESTADO DE SÂO PAULO, 2009).

Brazil is one of the few countries that grow acerola commercially (LOPES; PAIVA, 2008).

---

1 News provided by Mr. Osvaldo Dias, President of the Junqueirópolis Agricultural Association, at a Technical Meeting with CATI technicians from Junqueirópolis, researchers from the Alta Paulista and Alta Sorocabana Regional Poles / APTA, FATEC professors and ESALQ students, in Junqueirópolis, in June 2010.

But even so, world production is still distributed in Barbados, Colombia, Costa Rica, Cuba, the United States, Puerto Rico, French Guiana, the Philippines, Haiti, the Caribbean Sea Islands, Mexico, Peru, Suriname, Venezuela, Vietnam and some African countries (MANICA et al. 2003a).

As the main importers, the United States, the Netherlands, Japan, Germany, France, Puerto Rico, Barbados, Trinidad, Hungary, Saint Vincent, Korea and Israel import approximately 40% of the volume produced in the form of acerola fruit and products (CORRÊA et al. 2002; OLIVEIRA; SOARES FILHO; CUNHA, 1998).

Europe and the United States together consume around 2000 tons per year^{-1} , and Japan alone around 1500 tons per year^{-1} (IBRAF, 1995).

Japan began to demand acerola by creating various products: juice, soft drinks, acerola water, ice cream, yogurts, milk drinks, "Taff-man" type concentrates, chocolates, sweets, candies, jams, purees (BARBOZA; TAVARES; MELO, 1996).

In Germany, the annual consumption of acerola juice is approximately 40 liters per person, and it is generally used as blends in juices. In the Brazilian market, acerola is mainly sold as fresh fruit or frozen pulp, but it is also consumed as jelly, syrup, whole juice, ice cream, vitamin capsules, cosmetics and other products (CARDOSO; LOPES; ALMEIDA, 2003).

2. 2 Botanical aspects and description of the plant

In Brazil, the study of acerola began around 1955 at the Rural University of Pernambuco, due to the discovery of the high content of ascorbic acid (vitamin C) present in the fruit. The authors report that the Malpighiaceae family comprises more than 60 genera and 1000 species, the most economically important genus being *Malpighia,* which has more than 40 species (KLUGE; REZENDE, 2003).

Asenjo (1980), Araùjo and Minami (1994); Nogueira (1997); cited by Konrad (2002), demonstrated different botanical classifications in relation to the botanical species belonging to the acerola tree, being found in the literature as *Malpighia punicifolia* L., *Malpighia emarginata Sessé e Mocino ex DC.* and *Malpighia glabra* L. However, Oliveira et al. (2003 a) mention that analyses carried out by the Herbario de Linnaeus and other sources have shown that *M. glabra* and *Mpunicifolia* are a single species, unlike the acerola tree, which has small, tasteless fruits and a low quantity of juice.

According to Alves and Menezes (1995), the scientific name for the acerola tree, *Malpighia emarginata Sessé e Mocino ex DC.,* was adopted at a meeting of the International Board for Plant Genetic Resources in Rome, Italy (IBPGR, 1986), and was also cited by Itoo, Aiba and Ishihata (1990) in Japan. Mondin, Oliveira and Vieira (2010) evaluated the karyotype and confirmed that *M. emarginata* D.C. is a diploid species (2n = 20).

Kluge and Rezende (2003) define the aceroleira as a medium-sized evergreen shrub, with a crown diameter of 2 to 3 meters. It has a single, branched trunk and a dense crown made up of numerous woody branches, usually curving downwards. The seeds are small, ovoid, not albuminated, measuring 3 to 5 mm in length and 2 to 3 mm in width (ARAÙJO; MINAMI, 1994; GONZAGA NETO; SOARES, 1994; SIMAO, 1971).

The leaves are opposite, with a short petiole, oval to elliptical, varying in length from 2.5 to 9.0 cm and width from 1.2 to 6.0 cm, with a generally acute base and apex; they are entire, often wavy, dark green and shiny on the upper side and pale green and opaque on the underside. The flowers are hermaphrodite, arranged in small pedunculated axillary clusters of three to five perfect flowers, which are 1 to 2 cm in diameter. The color of the flowers, before fertilization and depending on the genotype, can be white, light pink, dark pink or violet (OLIVEIRA et al. 2003a).

The same authors define the fruit as a tri-pyrenoid drupe with a thin epicarp, a mesocarp with large, juicy cells and an endocarp made up of three elongated, lignified stones, varying in size, shape and weight. The shape of the fruit can be round, oval or subglobose and the weight can vary from 3 g to 16 g. The color of the flesh can be yellow, orange or red. The pulp represents, on average, 70 % to 80 % of the total weight of the fruit and can be acidic or subacidic, slightly sweet or sweet acidic. Malic acid plays a major role in the aroma of the fruit, giving it a fragrance similar to that of apples.

Acerola flowers have no nectar and the pollen grains are sticky and cannot be disseminated by the wind. Pollination depends on pollinating insects such as bees of the species *Centris dirrhoda*, Centris *tarsata, Centris aenea* and Centris *bicolor*. And pollen viability can vary from 10% to 90% depending on the genotype (FREITAS; PEREIRA, 2004; RAW, 1979 cited by OLIVEIRA et al. 2003a; SIQUEIRA et al. 2011).

2. 3 Food and pharmacological importance

Producers and the consumer market for acerola (Malpighia emarginata Sessé e Mocino ex

DC.) arose because of the high content of vitamins, especially C, and compounds in the fruit, such as antioxidants (ASENJO; MOSCOSO, 1950).

According to Habone and Williams (2000), fruits in general contain minerals, fibers, vitamins and polyphenols. Studies have shown that phenolic compounds, especially flavonoids such as anthocyanins, have the ability to capture free radicals, i.e. have antioxidant activity which is directly related to the prevention of cardiovascular problems and blood circulation (NESS; POWLES, 1997; STOCLET et al., 2004).

By measuring the oxidizing activity equivalent to Trolox (a synthetic antioxidant) and Vitamin C in various commercial fruits, acerola pulp showed a higher antioxidant capacity than mango, strawberry, grape, açai, guava, blackberry, soursop, passion fruit, cupuaçu and pineapple (KUSKOSKI et al. 2006).

It also has high levels of vitamin C (1000 to 4676 mg /100 g pulp), anthocyanins which have an antioxidant action, a good source of calcium and iron, thiamine, riboflavin and niacin, precursors of vitamin B; acerola also has a low caloric value, 33 kcal per 100 g of the edible part (ALVES; MENEZES; SILVA, 1995; LOPES; PAIVA, 2008; RITZINGER; RITZINGER, 2011; TACO, 2011).

According to Meira (1995), acerola has a high vitamin C content, and a lack of vitamin C in the human body weakens connective tissues and capillaries, causing bleeding of the skin, gums, joint pain and edema, which leads to scurvy.

Humans cannot synthesize vitamin C, although it is essential for the normal functioning of the body. In the body's vital processes, vitamin C plays a role in the formation of collagen, a protein substance that binds cells together and supports connective tissue, is an important growth activator and interferes in the metabolism of iron, glucose and other carbohydrates (MANICA et al. 2003a).

Carotenoids are important for human nutrition because they function as precursors of vitamin A, which maintains the integrity of skin tissues, vision, reproduction and growth (CAVALCANTE, 1991).

Moura et al. (2007), evaluated the fruit of acerola clones and found average values of 2.47 $\mu g.g^{-1}$ of β-carotene; and considered that acerola fruit has a satisfactory nutritional value for this vitamin.

The acerola has a good pulp yield, around 70%. The main commercially exploited food products derived from this fruit are frozen pasteurized pulp and pasteurized juice. Acerola can also be processed into various other products, such as nectar, jelly, freeze-dried products, preserves, liqueur, wine, ice cream, syrup, candy, as well as being added to other fruit juices ("blends") to enrich them with vitamin C. The pharmaceutical industry uses acerola mainly to produce vitamin C capsules and vitamin compounds (FOLEGATTI; MATSUURA, 2003). For export, the pulp of green and ripe fruit is used, as well as frozen, green and ripe fruit in natura (CARDOSO; LOPES; ALMEIDA, 2003).

2. 4 Varieties of acerola

White Flower: grown in the states of Parà, Bahia, Sergipe, Pernambuco and Minas Gerais, the plant is characterized by a low, airy and dispersed crown, which favours fruit harvesting. It is susceptible to branch breakage, which can be overcome by pruning.

It produces fruit with an average weight of 5 g, red in color, not very consistent, with a content of 7 °Brix, 1500 mg of ascorbic acid / 100 g of pulp. It produces around 20 kg of fruit between the ages of 8 and 20 months and bears fruit well during the rainy season.

Sertaneja: widely grown in Pernambuco, Bahia, Sergipe, Minas Gerais and Parà. Medium-sized plant (2.0 m to 2.5 m high), with a whitish-pink flower after anthesis, dark green leaves with little hairiness. Mature branches have a white pigmentation and the ripe fruits have a shiny dark red color.

The vitamin C content is greater than 1500 mg of ascorbic acid / 100 g of pulp, average weight of 4 to 5 g, firm pulp and yield of 100 Kg / plant / year in adult plants, when irrigation is used (RITZINGER; RITZINGER, 2011; RITZINGER; SOARES FILHO; OLIVEIRA, 2003).

Okinawa: one of the most popular varieties of aceroleira in the Brazilian states of Bahia, Pernambuco, Sergipe and Minas Gerais. Adult plants can produce over 40 kg of fruit per plant every year, which would result in approximately 16 t ha^{-1} (RITZINGER; RITZINGER, 2004, 2011).

This acerole tree is one of the most suitable cultivars for industrialization due to the fruit's compositional characteristics such as color, soluble solids content, vitamin C content and productivity. One of the goals for the coming years in Espirito Santo is to increase the area planted with the Okinawa and Sertaneja varieties from 40 to 150 hectares, and to increase average productivity to 40 tons ha- 1 using irrigation throughout the production area (INCAPER, 2011).

This genotype was introduced from Japan. It is characterized by a well-defined plant, with a high canopy, small diameter and little ventilation, branches with intense foliage and flowering, but the fruiting rate is low in isolated plantations; the early detachment of the fruit means that there is little use of it; it produces large, red-coloured fruit, with an average weight of around 7 g; soluble solids around 7 °Brix, high vitamin C content (over 2.200 mg of ascorbic acid /100 g of pulp), with excellent consistency; susceptible to pests and diseases common to the acerola tree in the conditions of Belém, PA; produces, between 8 and 20 months of age, around 18 Kg of fruit / plant. Information on its characteristics was obtained from plantations in a humid ecosystem (COOPERATIVA AGRÌCOLA MISTA DE TOMÉ-AÇÙ, 1994 cited by RITZINGER; SOARES FILHO; OLIVEIRA, 2003).

Maranhao (2010) found a content of 2232 mg of ascorbic acid / 100 g of pulp, and 60 $\mu g.g^{-1}$ and 5 $\mu g.g^{-1}$ of anthocyanins in the peel and pulp, respectively, at 25 days after anthesis, in the final period of fruit ripening of the Okinawa acerola. The author concluded that this acerola at the ripe stage stands out for having significant levels of these bioactive compounds such as carotenoids, anthocyanins, flavonois and ascorbic acid which make this fruit a promising source of antioxidant compounds whose cultivation should be encouraged.

Olivier: the cultivar has large fruits, averaging 29.7 mm in diameter and 24.2 mm in height, with an average weight of 10.4 g, ascorbic acid content of 2178.8 mg / 100g of pulp in unripe fruits and 1567.2 mg / 100g of pulp in ripe fruits, soluble solids of 9.92 °Brix and intense red color when ripe (KANNO; RIZZI; KAVATI, 2000).

These authors reported that the selection was carried out in 1995 at Fazenda Sâo Fortunato, owned by Francisco and Moacir Olivier in Junqueirópolis-SP, from an orchard probably planted by seed. The adult orchards achieved average yields of over 100 kg / plant / year. Another factor that favored the rapid expansion of the area cultivated with Olivier, according to the authors, was the high rooting capacity of semi-woody cuttings.

In the Alta Paulista region, the Olivier variety still guarantees the safety of acerola production, but the search for new cultivars adapted to the region's soil and climate conditions is essential to mitigate one of the biggest threats to local production, which is working with just one acerola variety (FREITAS; VILPOUX, 2007).

The BRS clones selected by EMBRAPA come from individual plants in open-pollinated acerola progenies. These progenies resulted from a selection made in the commercial orchard of the company Frucesa, in the municipality of Jaguaruana-CE (PAIVA, 2003b).

From then on, another trial was set up at Fazenda Frutacor Ltda. in Limoeiro do Norte-CE, and based on the results of production, plant morphology and physical and chemical characteristics of the fruit, four low-growing clones were selected and recommended: BRS 235 or Apodi, BRS 236 or Cereja, BRS 237 or Roxinha and BRS 238 or Frutacor (PAIVA et al. 2003).

BRS 235 or Apodi: has average values of 1.91 m for height, 4.05 m for crown diameter; 1260.9 mg ascorbic acid/100 g pulp, 6.4 °Brix, average fruit weight of 11.8 g and yield of 49.34 t ha^{-1} .

BRS 236 or Cereja: has average values of 2.19 m for height, 4.14 m for crown diameter; 1854.9 mg ascorbic acid/100 g pulp, 6.2 °Brix, average fruit weight of 7.56 g and yield of 47.19 t ha^{-1} .

BRS 237 or Roxinha: average height of 1.61 m, crown diameter of 3.57 m, 1193.9 mg ascorbic acid/100 g pulp, 8.3 °Brix, average fruit weight of 10.8 g and yield of 36.6 t ha 1.$^{-}$

BRS 238 or Frutacor: average height of 1.81m, crown diameter of 3.65m, 1656.1 mg ascorbic acid/100 g pulp, 8.3 °Brix, average fruit weight of 6.6 g and yield of 41.42 t ha 1.$^{-}$

In Cruz das Almas, the Apodi, Cereja, Frutacor, Okinawa and Sertaneja varieties had the highest levels of vitamin C, all above 1600 mg /100 g of pulp in a comparative study on fruit quality (RITZINGER; RITZINGER, 2011).

Waldy CATI 30: was obtained from seedlings produced by seeds in a Unit of the Coordination of Integral Assistance (CATI), where the planting was done on the property of Mrs. Waldy Hernandes in Jumirim-SP, which obtained good results. From then on, branches of the plant were collected for the multiplication of clones at CATI's Seedling Production Unit in Tietê-SP. The adult plant reaches a height of 4 m. The fruit is deep red in color, with 7.4 °Brix of soluble solids and 1493.9 mg of ascorbic acid/100 g of pulp (CATI, 1997).

2. 5 Propagation of the acerola tree

Propagation is a set of practices designed to perpetuate species in a controlled manner. Its aim is to increase the number of plants while ensuring that the essential agronomic characteristics of the cultivars are maintained (HOFFMANN; FACHINELLO; NACHTIGAL, 2005).

For the plant breeding researcher, the farmer and the industry, having full knowledge of the propagation method is fundamental, as it seeks to guarantee uniform and quality plantings of acerola (GOMES et al. 2000).

According to Lorenzi et al. (2006), the acerola tree is propagated by seeds, but the techniques of cuttings, grafting and micropropagation can also be used; and there are reports of propagation by cuttings and dives (KLUGE; REZENDE, 2003; MANICA et al. 2003b; SIMAO, 1998).

2. 5. 1 Propagation of the acerola tree by seed

The use of seeds to multiply acerola is mainly used to produce seedlings for genetic selection or rootstocks. The high heterozygosity favors the possibility of the plants formed showing variation in agronomic characters, and as the planting area increases, the generation of undesirable matrices from a genetic point of view increases, not serving as a recommendation for setting up commercial plantations (COUTO, 1997; NAKASONE; PAULL, 1998).

Propagation by seed is problematic for the acerola tree due to the lack of embryos, which can often be more than 50%. The lack of embryos in the seeds is due to possible incompatibility problems, which are greater when the plant is self-sown and lesser when different varieties are crossed. In this respect, the production of free-standing seedlings is more costly in terms of seed extraction, substrate and service time (ARAÙJO; MINAMI, 1994; COUTO, 1997; MUSSER; COUCEIRO; ALBUQUERQUE, 1987).

In addition to the significant absence of embryos in acerola seeds, Paiva et al. (1999) found germination rates ranging from 0% to 84.5% in over 19,000 seeds collected from 100 acerola trees selected from commercial orchards. This result showed that it is possible to obtain plants with seed production potential for rootstock production.

In Brazil, Oliveira et al. (2003b), report that the vast majority of commercial acerola plantations were generally established on the basis of seedlings obtained from seeds, resulting in great unevenness between plants, with negative repercussions on productivity and fruit quality. The use of vegetative propagation should therefore be preferred, as it allows for the multiplication (cloning) of individuals with superior agronomic characteristics, allowing for greater income for the producer.

It is important to note that seedlings produced from seeds do not produce early and start producing later than seedlings propagated by cuttings or grafting (RITZINGER; RITZINGER, 2003). According to EMBRAPA [2000?], plants obtained from seeds usually start producing from the second year after planting, even when they are irrigated in a semi-arid environment or grown in regions with a high rainfall.

2. 5. 2 Propagation of the acerola tree by cuttings

The use of cuttings and grafting as vegetative propagation techniques keeps the genetics of the mother plant unchanged and standardizes commercial orchards, making it easier for the grower to carry out daily cultivation. In the production of acerola by cuttings, the time needed to form the seedling is three to four months shorter than in the case of grafting (GONZAGA NETO; AMARAL; SAUERESSIG, 1996).

Simâo (1998) mentions that of all the propagation methods, cuttings seem to be the most suitable, as they guarantee the characteristics of the material to be propagated, but explains that the rooting and setting of cuttings can be low in the production of seedlings.

In terms of production costs, grafted seedlings are more economical than cuttings, as the process of cuttings requires the use of intermittent nebulization and plant regulators to achieve good plant setting efficiency in the nursery (RITZINGER; RITZINGER, 2003).

To stimulate or accelerate the rooting of acerola cuttings, Couto (1997) reports that producers use plant regulators such as Indolbutyric Acid, whose aim is to produce roots with a higher standard and ensure a stronger root system.

The cuttings can be treated with plant growth regulators such as indolbutyric acid at a concentration of 2000 to 2800 ppm, in order to accelerate the emission of roots (ALVES, et al. 1991; GONTIJO, et al. 2003).

Acerole woody cuttings treated with 1500 ppm of indolebutyric acid showed more significant results when compared to doses of 500 and 1000 ppm (SINGH; ATTRI, 2000).

However, Silva (2004) observed that the propagation of acerola by herbaceous cuttings is feasible, and showed 68% rooting, which is higher than woody cuttings, and that the application of indolbutyric acid or naphthalene acetic acid decreased the percentage of rooting for herbaceous cuttings, but increased it for woody cuttings. The time taken for acerola cuttings to form roots was more than 60 days.

Lima et al. (2006) indicated 10 cm long cuttings taken from the median portion as suitable material for producing acerola seedlings by cuttings. The cuttings were taken from progeny 91 of the genetic material collection of Embrapa Agroindustria Tropical in Pacajus-CE, with an average rooting rate of 63.33 % and significant results in terms of seedling vigor, such as the dry matter mass of the aerial part (2.10 g) at 60 days after planting, in the period from November 2000 to January 2001.

Ritzinger and Ritzinger (2011) state that it is essential to propagate the cuttings under reduced lighting (50% shading) and humidity saturation through intermittent misting or by wrapping the boxes or trays in transparent plastic bags. Rooting by cuttings can take between 40 and 60 days, with rooting efficiency depending on the genotype used.

Alves (1992) observed that acerola cuttings under intermittent nebulization with 70 to 80% humidity, and a substrate containing washed sand and vermiculite in a ratio of 4:1, resulted in 70% rooting and did not differ from cuttings with growth regulators.

Regardless of the dosage and use of plant regulators, Bezerra et al. (1991) found different values for the rooting of cuttings according to the time they were taken. Herbaceous cuttings taken in April had a higher rooting percentage (87.3%) than those taken in February (47.5%).

The best rooting results for acerola cuttings were 52.5% with herbaceous cuttings, followed by 27.1% for semi-woody cuttings and 6.78% for woody cuttings between April and June 1993 in Londrina-PR. In December of the same year, no significantly different results were observed in relation to the types of cuttings (GONZALEZ; PIPOLO; MALAGUIDO, 1994).

Nasser et al. (2010) propagated several acerola cultivars by cuttings using semi-woody cuttings collected in June 2010 in Adamantina-SP, and found an average rooting rate of 37.5% after 60 days of planting in an intermittent misting system and under a substrate of washed sand and without the use of a plant regulator.

Oliveira et al. (2003b), recommend collecting cuttings from pre-selected acerole trees that show good productivity and are free from pests and diseases. It is recommended to use young, vigorous, strong cuttings from terminal or herbaceous branches, 10 to 15 cm long, with two to four internodes and two pairs of leaves. The cuttings should be collected and planted immediately afterwards, burying the lower third to ensure a higher percentage of rooted cuttings. The author indicates that the substrate used to support the cuttings should be rich in nutrients if the option is to produce cuttings in a short period of time; and less nutrients in the substrate if the option is to produce cuttings over a longer period of time, around 60 days. To maintain humidity in the environment where the cuttings will be rooted, the author recommends an intermittent misting system.

It is also known that the plant propagated by cuttings provides an adventitious root system, which is therefore more superficial when compared to the plant obtained by grafting, which allows for more efficient exploration of the soil and the production of cuttings, and cuttings require an intermittent nebulization system and a structure to support the substrate for rooting (GONZAGA NETO; AMARAL; SAUERESSIG, 1996).

2. 5. 3 Propagation of the acerole tree by grafting

Modern fruit growing is based on vegetative propagation, i.e. grafting commercial varieties onto rootstocks, often obtained from seeds, such as citrus, avocado, cashew, mango, persimmon, etc (SIMÂO, 1998).

A plant propagated by grafting is basically made up of two parts: the scion and the rootstock. The scion is the part represented by a fragment of the plant, containing one or more buds, responsible for forming the aerial part of the new plant. The rootstock is the part responsible for forming the root system (NACHTIGAL; FACHINELLO; HOFFMANN, 2005).

The plant with the greatest potential for exploring the soil, in the event of any kind of restriction to root development, is the one propagated by grafting, as it has the pivoting root system of the rootstock (from seed) and the aerial part cloned from selected material (SOUZA; GOMES; SOUZA, 2003).

Grafting is the joining of two plants of the same variety or different varieties and species with possible interdependence between them, and makes it possible to obtain uniform plants in terms of genetics, as well as earlier production than plants propagated by seeds (HARTMANN et al. 1997; SIMÂO, 1998).

According to Gonzaga Neto and Soares (1994), Gonzaga Neto (1995), and Oliveira et al. (2003b), seedlings obtained by grafting develop a more vigorous root system, which consequently explores a greater volume of soil. In addition, the presence of the pivot root in the seedling obtained by grafting provides greater support for the plant in the soil, a favorable condition when planting orchards in regions subject to strong winds.

The main reason for using vegetative methods such as grafting to multiply plants is that this propagation consists of the mitotic division of cells, which means that the chromosome system and cytoplasm of the progenitor cell are completely duplicated. As a result, plants propagated by grafting carry all the genetic information of the parent plant. In the case of the acerola tree, grafting will be of great importance and indispensable use as soon as breeding work has a nematode-resistant clone (GONZAGA NETO, 1995).

Plant multiplication by grafting can also offer the following benefits: propagating plants that are difficult to grow by other methods, obtaining benefits from the rootstock, replacing established plant cultivars, recovering parts of plants damaged by low temperatures, diseases, winds, animals, agricultural equipment; studying viral diseases, plant development, physiological processes and combining cultivars or clones (HARTMANN et al., 1997; NACHTIGAL; FACHINELLO; HOFFMANN, 2005; NAKASONE; PAULL, 1998).

Grafting is a grafting method that consists of removing a portion of a branch, called a fork or scion, in the shape of a bevel or wedge, containing two or more buds, to be inserted into the rootstock or horse (NACHTIGAL; FACHINELLO; HOFFMANN, 2005).

Gonzaga Neto (1995) considers compatibility between scion and rootstock to be an important aspect, as the grafted plant can express its full potential in terms of productivity and longevity. The author states that in the acerola crop there is still not enough information, but observation is needed to avoid grafting plants with significant differences in size and different trunk diameters on plants of the same age.

Greater or lesser compatibility is closely related to the following factors: physiological, biological, tissue consistency, anatomical, size, vigor and sensitivity to viral diseases (SIMÂO, 1998).

Pompeu Júnior (2005), mentions that in citrus, the rootstock induces changes in growth, size, precocity of production, production, maturation and weight of the fruit, color of the peel and juice, content of sugars, acids and other components of the juice, permanence of the fruit on the plant and its preservation after harvest, pollen fertility, absorption, synthesis and use of nutrients, transpiration and chemical composition of the leaves, response to products of fruit and leaf abscission, tolerance to salinity, drought, cold, diseases and pests.

Regarding the setting of grafted acerola seedlings, Ritzinger and Ritzinger (2003), report that when produced under suitable conditions, they are highly efficient, being over 70%, but the authors do not inform the scion and rootstock used. Oliveira et al. (2003b) recommend grafting by full cleft grafting, but there are also recommendations for grafting by simple English grafting, lateral cleft grafting and normal T, inverted T and open window grafting (COUTO, 1997; GONZAGA NETO; AMARAL; SAUERESSIG, 1995; MANICA et al. 2003b; SIMAO, 1998; VIEIRA et al. 1997).

Grafted acerola seedlings, like cuttings, produce early, starting between 8 and 10 months after planting or in less than a year (OLIVEIRA et al., 2003b; RITZINGER; RITZINGER, 2003).

Even though there is still no genotype for rootstock in the acerola crop to control nematodes, it is believed that the results of acerola production in an area contaminated by this soil pest and which has been propagated by grafting is greater than a plant propagated by cuttings (GONZAGA NETO et al. 1999).

Commercial plantations of grafted acerola seedlings can be found in the Northeast region of Brazil, specifically in Ubajara, Ceará State. The company exports organic acerola all over the world and has a processing capacity of 85 tons per day^{-1} (FAZENDA NUTRILITE BRASIL, 2011).

The costs of producing grafted acerola seedlings were not disclosed, but EMBRAPA (2013) sells the seedlings at R$ 3.00 per unit. The Junqueirópolis-SP Agricultural Association, in partnership with the town hall, produces acerola seedlings by cuttings at a production cost of R$1.00, but sells them to members for R$0.50 and to non-members for R$2.50. The Coordenadoria de Assistência Tècnica Integral (CATI) is a São Paulo state institution responsible for rural extension, but they also sell seedlings and seeds of various agricultural species. CATI's Seed and Seedling Station in Lucélia-SP produces acerola seedlings by cuttings at a cost of R$2.50 and sells them for R$5.00 (verbal information).[2]

2 News provided by Agricultural Engineer Marcelo Rosa Melo from CATI's Seed Production Center in Lucélia in May 2013.

3. MATERIAL AND METHODS

3. 1 Climatic characterization of the experimental area

The experiment was conducted from January 10 to December 12, 2011 in the experimental nursery of the Alta Paulista Regional Center of the São Paulo Agency for Agribusiness Technology (APTA), linked to the São Paulo State Department of Agriculture and Supply, based in Adamantina-SP, whose geographical coordinates are: latitude 21° 40' S, longitude 51°08' W and altitude 400 m.

The region's climate is Cwa and classified as humid subtropical according to the Koppen classification, with hot, rainy summers and dry, mild winters (HERRERA et al. 1997). The average rainfall is 1283 mm and the average annual temperature is around 24 °C (CIIAGRO, 2013).

3. 2 Seedling formation

The propagation materials (seeds and forks) were taken from vigorous mother plants, 5 years old, with a planting spacing of 5.00m x 5.00m and propagated by cuttings, with the exception of Okinawa, which used grafted seedlings.

The plants installed in the active germplasm bank in the orchard of the experimental station itself had good fruit production and were free of pests and diseases.

The total training period for the seedlings was 11 months, with 48 days from sowing to pricking out the seedlings, 210 days from pricking out to grafting and 75 days from grafting to the final stage.

To sow the rootstocks, ripe fruit was collected from the field on January 10, 2011.

The seeds were then cleaned. This was done by manually removing the peel and pulp from the fruit by pressing them on a metal sieve with a diameter of approximately 60 cm and a mesh size of 4. The seeds were then washed under running water to remove any remaining pulp and peel that might have adhered to the seeds. The seeds were then sown in a sand germinator with a depth of between 3 and 4 cm and a spacing of 10 cm between rows.

3. 3 Cultivation

After sowing, a preventive control was carried out to prevent the seedlings from toppling over, caused by the fungus *Rhizoctonia solani*. For this, the fungicide Pencycuron

(Moncerem PM) was used, at a dose of 25 g i.a. per 100 L of water applied by watering over the sand. After emergence, the seedlings were transplanted into the containers when they were approximately 5 cm tall (48 days after sowing).

From the time the seedlings were pricked out until the final stage of the experiment, the seedlings were raised under a polypropylene screen at 50% shade and placed in plastic tubes suitable for fruit seedlings with a volume of approximately 300 mL. The substrate was Bioplant Prata HT® , whose main raw material is pine bark, and to a lesser extent coconut fiber, vermiculite and rice husk, as well as nutrients.

For each 25 kg bag of substrate, 250 g of the slow-release fertilizer Osmocote® in the 18-05-09 formulation was added.

To control the black flea, *Toxoptera citricidus* (Kirkaldy), foliar sprays were made of methomyl-based insecticide at a dose of 21.5 g a.i. per 100 L of water, and deltamethrin at a dose of 2500 g a.i. per 100 L of water, as necessary.

Irrigation was carried out daily by microsprinkling, and weeds were pulled out by hand as necessary.

During the pricking-out phase until grafting, the seedlings of all treatments received an application of 80 mg of Mono-Ammonium-Phosphate (MAP) fertilizer per seedling diluted in water (15 g in 10 liters of water), by watering the plants until the water ran off.

When the seedlings were being raised before grafting, four cuttings were made to keep them on a single stem.

The grafting process by full cleft grafting, as well as the other cultural treatments involved in the production of the acerola seedlings, followed the recommendations of Oliveira et al. (2003b).

On September 26, 2011, about 7 months after pricking out, the rootstocks were pruned at the apex to a cutting height of 15 cm (Figure 1).

Figure 1 - Cutting off the apical part of the acerola rootstock and keeping the leaves. Adamantina-SP, 2011.

Source: Author's own elaboration.

The forks used (semi-woody) were collected from the apical region of adult plants and were approximately 15 cm in size and similar in diameter to the rootstocks in the grafting region (Figure 2).

Figure 2 - Appearance of the Okinawa acerola forks used for grafting.

Adamantina-SP 2011

Source: Author's own elaboration.

The type of grafting used was full slit grafting, after which all the grafts were protected with a transparent plastic bag (40 x 28 cm), tying its lower end to the rootstock to protect the grafted area (Figure 3).

Figure 3 - Process of grafting by forking in a full cleft filled with aceroleira and protection with a plastic bag. Adamantina-SP, 2011.

Source: Author's own elaboration.

The existing leaves below the grafting point were kept (Figure 1). The protective bag was removed after the first leaves of the graft opened on October 17, 2011 (about 20 days after grafting).

3. 4 Treatments used

The treatments used were rootstocks:

1- Okinawa.

2- Olivier.

3- Apodi- BRS 235.

4- Cherry - BRS 236.

5- Roxinha - BRS 237.

6- Frutacor- BRS 238.

7- Waldy CATI 30.

The canopy used was the Okinawa cultivar.

3. 5 Experimental design

The experimental design was entirely randomized with 7 treatments (rootstocks) and 3 replications per treatment. Each experimental plot or repetition was made up of 9 seedlings, the useful plot being made up of the 7 central seedlings.

3.6 Variables analyzed

At 40, 55 and 75 days after grafting, the graft setting index was assessed. To do this, the surviving seedlings from each experimental plot were counted. This survival was characterized by the presence of shoots above the grafted area and the value was expressed as a percentage of setting.

During the same period, the initial development of the surviving seedlings was assessed using the following variables:

- height of the grafted seedlings: measured in cm using a graduated ruler. The ruler was positioned close to the seedling, starting from the neck to the last leaf insertion,
- stem diameter of the rootstock: measured in millimeters, using a digital caliper and carried out in the region of the seedling's neck,
- Graft stem diameter: measured in mm using a digital caliper and taken 1 cm above the grafting region,
- number of leaves issued per plant: determined by counting each leaf issued from each surviving seedling.
- number of sprouts emitted from the scion per plant: verified by counting each sprout emitted from the seedlings,
- dry matter mass of the aerial part, roots and total.

After assessing the variables mentioned above, the seedlings were removed from the containers and then washed in water to separate them from the substrate. Then the aerial part and the roots of each seedling were removed and separated using scissors, the cut being made in the region of the neck.

All the parts were placed in paper bags and dried in an oven with air circulation/renewal at 65°C until they reached a constant weight. Afterwards, the materials were removed from the oven and the dry matter masses of the aerial part, roots and total were measured in grams using a Marconi electronic balance® .

A period of 75 days after grafting was considered adequate for the acerola seedlings to be ready for acclimatization and planting in the field.

3. 7 Statistical analysis

The data obtained was subjected to an analysis of variance and when significance was found by the F test, the means of the treatments were compared using the Duncan test at a significance level of 5%. The statistical program used was SAS (S.A.S., 1999).

4. RESULTS AND DISCUSSION

4. 1 graft setting index

Table 1 shows the graft setting data. Although there was no significant difference, all the rootstocks used showed good graft set. When the Apodi rootstock was used, an average of 95.24% of the grafts took hold by the end of the period analyzed. The Okinawa and Olivier cultivars used as rootstocks also showed satisfactory values, with more than 85% of the grafts sticking.

Table 1- Average values of grafting percentage at 40, 55 and 75 days after grafting the Okinawa acerole tree onto different rootstocks. Adamantina-SP, 2011.

Rootstocks	Graft setting (%)		
	40	55	75
		Days after grafting	
Okinawa	85,71	85,71	85,71
Olivier	80,95	85,71	85,71
Apodi	95,24	95,24	95,24
Cherry	90,48	90,48	90,48
Roxinha	80,95	80,95	80.95
Frutacor	85,71	90,47	90,47
Waldy CATI 30	61,90	61,90	61,90
Average	82,99	84,35	84,35
F	1,33	1,41	1,41
C.V. (%)	19,15	18,84	18,84

Source: Author's research data.

The Waldy CATI 30 cultivar showed the lowest setting (61.90 %) at 75 days after grafting. This is probably due to a biological factor such as different physiological characteristics, or anatomical affinity where the scion and rootstock may have cells with different size, shape and consistency; the cambial tissues are not joined to form a continuous connection

(NACHTIGAL; FACHINELLO; HOFFMANN, 2005).

In plant grafting in general, Hartmann et al. (1997) show that there are differences between species and cultivars in terms of how well the various grafting methods take hold and that the variation is related to the ability to produce callus from parenchyma, which is essential for the success of the union.

For the other treatments, the average values were higher than 80%, showing a high rate of graft take when grafting by top grafting in a full slit with transparent plastic bag protection and under shaded roof conditions. Oliveira et al. (2003b), indicate that when using grafting in aceroleira by full cleft grafting, and under suitable conditions, the expected graft setting will be close to 80%.

Looking at Table 1, it can be seen that the average of all the treatments at 55 and 75 days resulted in 84.35 % of grafted acerola seedlings taking hold, and was higher than the value found by Gonzaga Neto, Amaral and Saueressig (1996), who found an average setting index of 73.3 % with 8-month-old seedlings, using the same grafting method, under greenhouse conditions in Petrolina-PE. However, the authors did not define the cultivar used in the experiment.

In a study without information on the variety evaluated for rootstock and scion, Araùjo et al. (1994), found a setting index of up to 82.5% in seedlings grafted with aceroleira by top grafting in a full slit at 6 months of age, in winter in Recife-PE, from seeds through mature fruit harvested from plants in production.

At the Itambé Experimental Station, in the state of Pernambuco, Pedrosa et al. (1994) found a 65.5% yield from September to December 1993, when they used full slit forking with rootstocks and forks made up of seeds from various unidentified acerola trees.

From all of the above results, it is important to highlight the high grafting rate of 85.71% for Okinawa as a rootstock, given the difficulties in propagating it by seed and by cuttings. In this respect, Simplicio et al. (1994) and Souza Júnior et al. (1994), assessing the presence of embryos in acerola seeds, found that Okinawa seeds had between 70.45 % and 74.34 % of embryos, 7.01 % to 19.32 % of atrophied embryos and 5.81 % to 22.54 % of seeds with normal embryos.

Gonzaga Neto et al. (1999) reported that the rooting of the Okinawa variety propagated by cuttings, without the use of plant regulators, resulted in low take-up, while Nasser et al. (2010) propagated the same cultivar, and after 60 days of cuttings found 46.6% of rooted cuttings, with material collected in a cold, dry season, and without the use of any regulator.

4. 2 Initial development

4. 2. 1 Height of grafted seedlings

With regard to the height of the grafted seedlings (Table 2), it can be seen that at 55 days after grafting, Okinawa on Roxinha rootstock did not differ from Okinawa, Apodi, Cereja and Frutacor, but at 75 days there was no significant difference between all the rootstocks used (Figure 4).

At the end of the analysis period, the treatments ranged from 30.19 cm when using Frutacor rootstock to 33.03 cm for Apodi rootstock.

Table 2- Average height values (cm), at 40, 55 and 75 days after grafting, of the Okinawa acerole tree on different rootstocks. Adamantina-SP, 2011.

Rootstocks	Height of Plant (cm)		
	40	55	75
Days after grafting			
Okinawa	29,61	30,58 abc*	31,12
Olivier	29,04	29.64 bc	30,35
Apodi	29,56	30.86 ab	33,03
Cherry	29,27	30.15 abc	31,20
Roxinha	30,51	32,08 a	32,35
Frutacor	28,53	30,16 abc	30,19
Waldy CATI 30	26,40	27,68 c	30,22

Average	28,99	30,16	31,21
F	2,20	3,18	1,09
C.V. (%)	5,21	5,15	5,89

*Means followed by the same lower-case letter in the columns do not differ by Duncan's test at 5%.

Source: Author's research data.

Figure 4 - Visual appearance of Okinawa acerola seedlings grafted onto different rootstocks at 40 (A), 55 (B) and 75 (C) days after grafting. Adamantina-SP, 2011.

Source: Author's own elaboration.

4. 2. 2 Stem diameters of rootstock and scion

For the rootstock and scion stem diameters (Table 3), the results showed no significant difference between the treatments during the periods analyzed, demonstrating that there was no influence from the rootstock. However, there was a numerically higher value (10.16 mm) for the rootstock of the Waldy CATI 30 cultivar at 75 days after grafting, probably due to the greater mortality of the seedlings grafted onto this treatment, which provided greater spacing between the plants in the plot.

Another factor that occurred in the treatment with the Waldy CATI 30 rootstock was a possible incompatibility perceived in the seedlings. Simao (1998) mentions that the scion or graft can influence the rootstock, changing the development and distribution of the root system in the soil, and these changes are difficult to measure.

Table 3 - Average rootstock and scion stem diameter values in mm at 40, 55 and 75 days after grafting, of Okinawa acerola seedlings on different rootstocks. Adamantina-SP, 2011.

	Rootstock stem diameter (mm)			Graft stem diameter (mm)		
Rootstocks	40	55	75	40	55	75
	Days after grafting					
Okinawa	5,66	5,71	7,69	3,89	3,96	4,59
Olivier	5,92	5,93	8,08	3,95	4,08	4,61
Apodi	6,40	6,46	7,65	4,08	4,23	5,00
Cherry	5,73	5,75	7,41	4,01	4,08	5,01
Roxinha	6,41	6,58	8,29	4,06	4,10	5,02
Frutacor	5,79	5,87	7,43	3,81	3,87	4,60
Waldy CATI 30	5,48	5,52	10,16	3,52	3,53	4,62
Average	5,91	5,97	8,10	3,90	3,98	4,78
F	0,63	0,87	1,25	0,68	0,55	0,37
C.V. (%)	13,29	12,83	18,42	12,25	11,86	12,99

Source: Author's research data.

The author exemplifies this phenomenon in citrus, when the rootstock *Poncirus trifoliata is* used in some citrus species. The horse has a trunk with a larger diameter when compared to its own canopy. What can also happen, according to Simao (1998), is that when the scion is vigorous, it induces greater development in the rootstock.

4. 2. 3 Number of leaves per plant

Initial development at 40 days after grafting, when the rootstock used was Okinawa, showed an average of 15.91 leaves and was significantly lower than all the other treatments (Table 4). At 55 days after grafting, the Waldy CATI 30 rootstock had 32.74 leaves, a statistically low number compared to the other rootstocks, while the Cherry and Roxinha treatments were significantly superior with 45.47 and 45.72 leaves respectively.

Table 4 - Average values for the number of leaves at 40, 55 and 75 days after grafting, of seedlings of aceroleira cv. Okinawa on different rootstocks. Adamantina-SP, 2011.

Number of leaves per plant			
Rootstocks	40	55	75
		Days after grafting	
Okinawa	15,91 b*	35.67 bc	49,09
Olivier	22.77 ab	41.75 abc	56,34
Apodi	28,37 a	44.61 ab	61,70
Cherry	28,52 a	45,47 a	58,62
Roxinha	29,01 a	45,72 a	55,76
Frutacor	20.56 ab	38.63 abc	50,47
Waldy CATI 30	21.89 ab	32,74 c	46,76
Average	23,86	40,66	54,11
F	3,01	3,25	2,48
C.V. (%)	20,75	12,10	11,07

*Means followed by the same small letter in the columns do not differ by Duncan's test at 5%.

Source: Author's research data.

At 75 days after grafting, when the seedlings could be used for planting, there was no statistical difference for this variable, with an average value of 54.11 leaves.

The results obtained for the number of leaves on the scion of the Okinawa cv. were higher than those found by Vieira et al. (1997), in a study using grafting on acerole seedlings of an unknown variety, where they found higher values at 60 days after grafting, using lateral slit grafting, with an average value of 15.45 leaves.

4. 2. 4 Number of shoots per plant

In terms of the number of shoots emitted, the Okinawa cv. acerole seedlings grafted onto the Roxinha, Frutacor and Waldy CATI 30 cultivars initially showed the lowest number of shoots emitted (Table 5). The Apodi rootstock at 55 days after grafting showed significant superiority in the number of shoots emitted per plant (14.22), and even though it resulted in the highest numerical value, with an average of 15.87 shoots emitted at 75 days after grafting, there was no significant difference. This shows that initial development was lower for most of the rootstocks used, but then the treatments were equal.

The results were higher than those found by Vieira et al. (1997), when they worked with grafting by forking on acerola seedlings of an unknown variety, and found the highest values at 60 days after grafting, using lateral slit grafting, with an average value of 4.31 shoots emitted per plant.

Table 5 - Average values for the number of shoots emitted at 40, 55 and 75 days after grafting, from seedlings of aceroleira cv. Okinawa on different rootstocks. Adamantina-SP, 2011.

Number of shoots per plant			
Rootstocks	40	55	75
		Days after grafting	
Okinawa	11.41 bc	12.29 abc	13,49
Olivier	11.91 abc	13.53 ab	14,30
Apodi	13.05 ab	14,22 a	15,87
Cherry	13,54 a	13.77 ab	13,95
Roxinha	10,87 c	11.07 bc	12,54
Frutacor	11,03 c	11.12bc	12,43
Waldy CATI 30	9,42 c	9,81 c	13,29
Average	11,60	12,26	13,70

F	4,01	3,14	2,57
C.V. (%)	8,55	12,46	9,27

*Means followed by the same lower-case letter in the columns do not differ by Duncan's test at 5%.

Source: Author's research data.

4 . 2. 5 Dry matter mass of aerial part, roots and total

Table 6 shows no significant difference between the treatments when analyzing the dry matter mass of the aerial part, roots and total, but there were higher values for the Apodi and Roxinha rootstocks, especially for the dry matter mass of the roots, with 6.48 g and 6.38 g, respectively. The Waldy CATI 30 rootstock showed the lowest values for dry matter mass of the aerial part (5.99 g), roots (5.55 g), and total (11.55 g). In the other treatments, Olivier, Cereja and Frutacor, the values were very close, ranging from 12.10 g to 12.74 g.

The values for dry matter mass of the acerola seedlings propagated by grafting in Table 6 are much higher than those of studies that analyzed this variable in acerola seedlings propagated by cuttings. Silva (2004), working with the rooting of acerola cuttings using different doses of plant regulators, found values of 1.16g for the mass of dry matter of the aerial part and 0.050g for the mass of dry matter of the roots at 60 days after the cuttings and with intermittent nebulization.

Under the same conditions, and using 10 cm long cuttings taken from the median portion of branches of progeny 91, the highest average values found by Lima et al. (2006) were 2.10 g for the dry matter mass of the aerial part, and 0.66 g for the dry matter mass of the roots.

Gonzaga Neto and Soares (1994), Nakasone and Paull (1998) and Ritzinger and Ritzinger (2003) emphasized that even though cuttings provide a faster way of obtaining acerola seedlings when compared to seedlings grafted onto seed-formed rootstocks, the latter has some advantages that should be taken into account, such as having a more vigorous, pivoting root system, penetrating deeper and tolerating water deficit for longer.

Table 6 - Average values of aerial dry matter mass (MMSPA), in g root dry matter mass (MMSR) and total dry matter mass (MMST) at 75 days after grafting, of acerole seedlings cv. Okinawa on different rootstocks. Adamantina-SP, 2011.

Rootstocks	MMSPA	MMSR	MMSTOTAL
		(g)	
Okinawa	6,17	5,61	11,77
Olivier	6,63	5,61	12,25
Apodi	6,77	6,48	13,25
Cherry	6,59	6,14	12,74
Roxinha	6,91	6,38	13,29
Frutacor	6,28	5,83	12,10
Waldy CATI 30	5,99	5,55	11,55
AVERAGE	6,48	5,94	12,42
F	0,48	1,07	0,67
CV (%)	13,06	10,96	11,74

Source: Author's research data.

5. FINAL CONSIDERATIONS

Under the conditions in which the experiment was conducted and according to the results obtained, it can be considered that:

The plants that were grafted in this work need to be evaluated at field level, checking their development during planting, post-planting, training and production. In this way, the viability of each material will be analyzed, as well as the possible influence of the rootstock on the agronomic characteristics.

In the productive phase, to verify productivity, vigor and evaluate the characteristics of acerola fruit and the composition of its pulp, and whether or not grafting modifies the quality of the fruit.

Due to the few existing studies on the propagation of acerola by grafting, other properly identified rootstock and scion genotypes should be used for evaluation and later to expand agronomic information on the subject, with the aim of studying productivity gains or selecting promising genetic material to control phytonematoids, which are considered to be important soil pests for the acerola crop.

There is a need for research into propagation using lateral grafting and other grafting methods, as well as studying the use of garter grafting at other times of the year.

A survey is needed of the production costs of grafted seedlings and their comparison with seedlings propagated by cuttings.

6. CONCLUSIONS

Based on the results obtained, and the conditions under which the work was carried out, it can be concluded that:

a) Due to the high percentage of grafting success on the different rootstocks, this technique proved to be viable for the crop;

b) There were no differences between the different scion/rootstock combinations in terms of the initial development of the seedlings, as assessed by plant height, scion and rootstock stem diameter, number of leaves and shoots produced, as well as the accumulation of dry matter mass in the aerial part and root.

c) The total time from sowing to the final stage of production of the grafted acerola seedlings was 11 months, and at 75 days after grafting they were ready for planting in the field.

7. REFERENCES

ADRIANO, E.; LEONEL, S.; EVANGELISTA, R. M. Fruits quality of acerola cv. Olivier at two stages of ripening. **Revista Brasileira de Fruticultura**, Jaboticabal, n. Especial, p. 541-545, 2011.

ALVES, R. E. Growing acerola. In: DONADIO, L. C.; MARTINS, A. B. G. (Orgs.). **Tropical fruit growing**. Jaboticabal: FUNEP, 1992. p. 15-37.

ALVES, R. E. et al. Acerola. In: CRISÓSTOMO, L. A.; NAUMOV, A. (Orgs.).

Fertilizing for high productivity and quality: tropical fruit trees in Brazil. Fortaleza: Embrapa Agroindùstria Tropical, 2009. p. 13-30.

ALVES, R. E. et al. Contribution to the study of acerola culture - I: effects of IBA and sucrose on the rooting of cuttings. **Revista Brasileira de Fruticultura**, Jaboticabal, v. 13, n. 2. p. 19-26, 1991

ALVES, R. E.; MENEZES, J. B. Botànica da aceroleira. In: SÂO JOSÉ, A. R.;

ALVES, R. E. (Eds.). **Acerola in Brazil**: production and market. Vitória da Conquista: State University of Southwest Bahia,1995. p. 7-14.

ALVES, R. E.; MENEZES, J. B.; SILVA, S. M. Harvest and post-harvest of acerola. In: SÂO JOSÉ, A. R.; ALVES, R. E. (Eds.). **Acerola in Brazil**: production and market. Vitória da Conquista: Universidade Estadual do Sudoeste da Bahia,1995. p. 77-89.

ANDRADE, P. F. S. **Anàlise da conjuntura agropecuâria, safra 2011/12 - Fruticultura.** Curitiba: Paranà Department of Agriculture and Supply, Department of Rural Economy, 2012. Available at:

<www.agricultura.pr.gov.br/arquivos/File/deral/Prognosticos/fruticultura_2011_12.p df.> Accessed: May 7, 2013

ARAÙJO, E. L. et al. setting index in grafted seedlings of acerola (Malpighia glabra), in two seasons and two ages of the rootstock. In: CONGRESSO BRASILEIRO DE FRUTICULTURA, 13, 1994, Salvador. **Proceedings...** Salvador: Sociedade Brasileira de Fruticultura, 1994. p. 66-67.

ARAÙJO, P. S. R.; MINAMI, K. **Acerola.**Campinas: Fundaçâo Cargill, 1994. 81 p.

ASENJO, C. F.; FREIRE de GUZMAN, A. R. The high ascorbic acid content of the West Indian Cherry. **Science**, Washington, v. 103, p. 219, 1946.

ASENJO, C. F.; MOSCOSO, C. G. Ascorbic acid content and other characteristics of the West Indian Cherry. **Food Research**, Chicago, v.15, p.103-106, 1950.

BARBOZA, S. B. S. C.; TAVARES, E. D.; MELO, M. B. Origin and importance. In: (Eds.). **Instructions for growing acerola**. Aracaju: Embrapa Tabuleiros

Coastal, 1996. 42 p. (Technical Circular, n. 6). Available at: <http://ainfo.cnptia.embrapa.br/digital/bitstream/item/44461/1/CPATC- DOCUMENTOS-6-INSTRUCOES-PARA-O-CULTIVO-DA-ACEROLA-FL- 13124.pdf>. Accessed on: May 29, 2013.

BEZERRA, F. E. J. et al. Effect of the size of herbaceous cuttings and indolebutyric acid on the rooting of acerola (*Malpighia glabra* L.) at two staking times. **Revista Brasileira de Fruticultura,** Jaboticabal, v. 13, n. 3, p. 157-163, 1991.

CARDOSO, C. E. L.; LOPES, R. L.; ALMEIDA, C. O. Economic aspects. In: RITZINGER, R.; KOBAYASHI, A. K; OLIVEIRA, J. R.P (Eds.). **The cultivation of acerola**. Cruz das Almas: Embrapa Mandioca e Fruticultura, 2003. p. 185-198.

COORDENADORIA DE ASSISTENCIA TÈCNICA INTEGRAL - CATI. **CATI launches new acerola:** cultivar "Waldy" CATI 30. Sao Paulo: CATI, 1997. 3 p.

cooRDENATIoN oF iNTEGRAL tEchNIC ASSISTANCE - cATi.

Census survey of agricultural production units in the state of São Paulo (LUPA) 2007/2008. Sao Paulo: SAA / CATI / IEA, 2008. Available at: <http://www.cati.sp.gov.br/>. Accessed on: September 8, 2012.

CAVALCANTE, M. L. **Carotenoid composition and vitamin A value in pitanga (*Eugenia uniflora*) and acerola (*Malpighia glabra*)**. 1991. 69 f. Dissertation (Master's Degree in Food Technology) - Federal University of Rio de Janeiro, Rio de Janeiro, 1991.

INTEGRATED AGROMETEOROLOGICAL INFORMATION CENTER - CIIAGRO. **Water balance by location:** Adamantina. Sao Paulo: SAA / IAC / CIIAGRO, 2013. Available at:

<http://www.ciiagro.sp.gov.br/ciiagroonline/Listagens/BH/LBalancoHidricoLocal.as p>. Accessed on: April 16, 2013.

CORRÊA, F. L. O. et al. Nutrient accumulation in acerola seedlings fertilized with phosphorus and zinc. **Revista Brasileira de Fruticultura**, Jaboticabal, v. 24, n. 3, p. 765-769, 2002.

CoUCEIRo, E. M. **Extension course on acerola cultivation**. Recife: Federal Rural University of Pernambuco, 1985. 45 p.

COUTO, F. A. A. **Production of acerola:** manual. Viçosa: CPT, 1997. 47 p.

BRAZILIAN AGRICULTURAL RESEARCH Corporation - EMBRAPA.

Questions and answers: acerola. Cruz das Almas: Embrapa Cassava and Fruit Growing, [2000?]. Available at:

<http://www.cnpmf.embrapa.br/index.php?p=perguntas_e_respostas-acerola.php>. Accessed on: May 30, 2013.

BRAZILIAN AGRICULTURAL SEEDLING COMPANY - EMBRAPA. **Sale of seedlings.** Cruz das Almas: Embrapa Cassava and Fruit Growing, 2013.

Available at: <http://www.cnpmf.embrapa.br/mdex.php?p=servicos-venda_de_mudas.php>. Accessed on: May 7, 2013.

FOOD AND AGRiCuLTuRE ORGANiZATiON OF THE uNiTED NATiONs -

FAO. **Food and Agricultural commodities production (Countries by**

Rome: FAO, 2011. Available at: <faostat.fao.org/site/339/default.aspx>. Accessed on: May 29, 2013.

NuTRiLiTE BRAsiL FAzENDA. **Largest exporter of organic acerola in the world**. Ubajara: Fazenda Nutrilite Brasil, 2011. Available at: <http://www.youtube.com/watch?v=cwAuzV8R5VY>. Accessed on: April 24, 2013.

FOLEGATTi, M. i. s.; MATsUURA, F. C. A. U. products. in: RiTZiNGER, R.;

KOBAYAsHi, A. K.; OLiVEiRA, J. R. p. (Eds.). **The cultivation of acerola**. Cruz das Almas: Embrapa Mandioca e Fruticultura, 2003. p. 164-184.

FREiTAs, B. M.; pEREiRA, J. O. p. Crop consortium to improve pollination: can West indian Cherry (*Malpighia emarginata)* attract *Centris* bees to pollinate cashew (*Anacardium occidentals)*? in: ________________ . (Eds.). **Solytary bees, conservation, rearing and management for pollination.** Fortaleza: Universidade Federal do Cearà, 2004. p. 193-201.

Available at: <http://www.webbee.org.br/bpi/solitary/livro_04.pdf>. Accessed on: May 31, 2013.

FREITAS, C. G.; VILPOUX, O. F. Associativism and solidarity economy in acerola production in the municipality of Junqueirópolis-SP. In: ENCONTRO INTERNACIONAL DE ECONOMIA SOLIDARIA, 5., 2007, Sao Paulo. **Electronic proceedings...** Sao Paulo: USP, 2007. Available at: <http://sites.poli.usp.br/p/augusto.neiva/nesol/Publicacoes/V%20Encontro/Artigos/Principios/PRI-05.pdf>. Accessed on: May 30, 2013.

GOMES, J. E. et al. Rooting of herbaceous cuttings of acerola genotypes in an intermittent misting chamber treated with indolbutyric acid at two different times.

Revista Brasileira de Fruticultura, Jaboticabal, v. 22, n. 3, p. 407-412, 2000.

GONTIJO, T. C. A. et al. Rooting of different types of acerola cuttings using indolbutyric acid. **Revista Brasileira de Fruticultura**, Jaboticabal, v. 25, n. 2, p. 290-292, 2003.

GONZAGA NETO, L. et al. **A cultura da acerola**. 2. ed., rev. e aum. Brasilia-DF: Embrapa Produçao de Informaçao; Petrolina: Embrapa Semi-Arido, 1999. 110 p.

GONZAGA NETO, L. Vegetative propagation: grafting in acerola. In: SÂO JOSÉ, A. B.; ALVES, R. E. **Acerola no Brasil:** produção e mercado. Vitória da Conquista: State University of Southwest Bahia, 1995. p. 42-46.

GONZAGA NETO, L.; AMARAL, M. G.; SAUERESSIG, M. E. Grafting by forking and budding in acerola under glass. **Pesquisa Agropecuâria Brasileira**, Brasilia-DF, v. 31, n. 9, p. 635-638, 1996.

GONZAGA NETO, L.; AMARAL, M. G.; SAUERESSIG, M. E. Vegetative propagation in acerola (*Malpighia glabra* L.) I. Nursery seedling production.

Revista Brasileira de Fruticultura, Jaboticabal, v. 17, n. 3, p. 93-96, 1995.

GONZAGA NETO, L.; SOARES, J.M. **Acerola para exportação:** aspectos técnicos da produção. Brasilia-DF: Embrapa-SPI, 1994. 43 p. (FRUPEX, 10).

GONZALEZ, M. G. N.; PIPOLO, V. C.; MALAGUIDO, A. B. Influence of physical consistency on the rooting of acerola cuttings (Malpighia glabra L.). In: CONGRESSO BRASILEIRO DE FRUTICULTURA, 13, 1994, Salvador. **Proceedings...** Salvador:

Sociedade Brasileira de Fruticultura, 1994. v. 1, p. 77.

HARBONE, J. B.; WILLIAMS, C. A. Advances in flavonoid research since 1992. **Phytochemistry**, New York, v. 52, p. 481-504, 2000.

HARTMANN, H. T. et al. The biology of grafting. In: HARTMANN, H. T.;

KESTER, D. E.; DAVIES JR., F. T.; GENEVE, R. L. (Eds.). **Plant propagation:** principles and practices. New Jersey: Prentice-Hall, 1997. p. 392-436.

HERRERA, O. M. et al. Grouping of climatological stations located in the State of Sao Paulo, using multivariate analysis. **Engenharia Agricola**, Jaboticabal, v. 16, n. 3, p. 34-42, 1997.

HOFFMANN, A.; FACHINELLO, J. C.; NACHTIGAL, J. C. Forms of fruit plant propagation. In: (Ed.). **Propagation of fruit plants**. Brasilia- DF: Embrapa Informaçao Tecnológica, 2005. p. 13-44.

BRAZILIAN INSTITUTE OF GEOGRAPHY AND STATISTICS - IBGE. **Production and area in agricultural establishments with more than 50 existing feet.** Rio de Janeiro: IBGE, 2006. Available at:

<http://www.sidra.ibge.gov.br/bda/agric/default.asp?t=2&z=t&o=11&u1=1&u2=1&u3=1&u4=1&u5=1&u6=1>. Accessed on: May 29, 2013.

INTERNATIONAL BOARD PLANT GENETIC RESOURCES - IBPGR.

Malpighia emarginata (Acerola). In: ______. **Genetic resources of tropical and subtropical fruits and nuts (excluding musa)**. Rome: IBPGR, 1986. p.52-54.

BRAZILIAN FRUIT Institute - IBRAF. **Acerola**. Sao Paulo: IBRAF, 1995. 62 p. (Soluçao fruta a fruta, 2).

CAPIXABA INSTITUTE FOR RESEARCH, TECHNICAL ASSISTANCE AND RURAL EXTENSION - INCAPER. **Incaper delivers more than 5,000 acerola seedlings to producers in Piùma.** Vitória: INCAPER, 2011. Available at: <http://www.seag.es.gov.br/?p=22617>. Accessed on: May 30, 2013.

ITOO, S.; AIBA, M.; ISHIHATA, K. Comparison of ascorbic acid content in acerola fruit from different regions depending on degree of maturity, and its stability by processing. **NipponShokukin KogyoGakkaishi**, Isawa-cho, v. 37, n. 9, p. 726-729, 1990.

JANICK, J.; PAULL, R. Malpighiaceae. In: ______(Eds.).**The encyclopedia of fruit and nuts**. Wallingford: CABI, 2008, p. 459-466.

KANNO, O. Y.; RIZZI, L. C.; KAVATI, R. Acerola Olivier. In: DONADIO, L. C. (Ed.). **New Brazilian fruit varieties.** Jaboticabal: Sociedade Brasileira de Fruticultura, 2000. p. 24-25.

KAWATI, R. Research and extension on acerola cultivation in the State of São Paulo. In: SÂO JOSÉ, A. B.; ALVES, R. E. **Acerola no Brasil:** produção e mercado.

Vitória da Conquista: State University of Southwest Bahia, 1995. p. 149-154.

KLUGE, R. A.; REZENDE, G. O. Aceroleira (*Malpighia* sp.). In: CASTRO, P. R. C.; KLUGE, R. A. (Coords.). **Ecophysiology of fruit trees:** avocado, acerola, apple, pear and grapevine. Piracicaba: Agronômica Ceres, 2003. p. 25-43.

KONRAD, M. **Effect of spot irrigation systems on the production and quality of acerola (*Malpighia emarginata* DC) in the Nova Alta Paulista region.** 2002. 119 f. Dissertation (Master's Degree in Agronomy) - Faculty of Engineering, Paulista State University, Ilha Solteira, 2002.

KUSKOSKI, E. M. et al. Wild tropical fruits and frozen fruit pulps: antioxidant activity, polyphenols and anthocyanins. **Ciência Rural**, Santa Maria, v. 36, n. 4, p. 1283-1287, 2006. Available at: <http://www.scielo.br/pdf/cr/v36n4/a37v36n4.pdf>. Accessed on: May 30, 2013.

LIMA, R. L. S. et al. Length of cuttings and part of the branch in the formation of acerola seedlings. **Revista Brasileira de Fruticultura**, Jaboticabal, v. 28. n. 1. p. 83-86, 2006.

LOPES, R.; PAIVA, J. R. Aceroleira. In: BRUCKNER, C. H. (Ed.). **Improvement of tropical fruit trees.** Viçosa: Federal University of Viçosa, 2008. p. 63-100.

LORENZI, H. et al. **Cultivated Brazilian and exotic fruits**: (for fresh consumption). Sao Paulo: Instituto Plantarum de Estudos da Flora, 2006. 640 p.

MANICA, I. et al. Economic importance. In: MANICA, I. et al. (Ed.). **Acerola:** production technology, post-harvest, freezing, exports, markets. Porto Alegre: Cinco Continentes, 2003a. p. 9-30.

MANICA, I. et al. Seedling production. In: MANICA, I. et al. (Ed.). **Acerola:** production technology, post-harvest, freezing, exports, markets. Porto Alegre: Cinco Continentes, 2003b. p. 89-133.

MARANHAO, C. M. C. **Physical, physicochemical and chemical characterization of the fruit of the acerola tree (*Malpighia emarginata* DC), Okinawa variety, during its development**. 2010. 73 f. Dissertation (Master's Degree in Food Science and Technology) - Federal University of Paraiba, Joâo Pessoa, 2010. Available at: <http://bdtd.biblioteca.ufpb.br/tde_busca/arquivo.php?codArquivo=917>. Accessed on: May 30, 2013.

MARINO NETTO, L. **Acerola:** the tropical cherry. Sâo Paulo: Nobel, 1986. 94 p. MEIRA, M. O. B. Vitamin C and its relationship with health. In: SAO JOSÉ, A. B.;

ALVES, R. E. **Acerola no Brasil:** produção e mercado. Vitória da Conquista: State University of Southwest Bahia, 1995. p. 1- 3.

MONDIN, M.; OLIVEIRA, C. A.; VIEIRA, M. L. C. Karyotipe characterization of *Malpighia emarginata* (Malpighiaceae). **Revista Brasileira de Fruticultura**, Jaboticabal, v. 32, n. 2, p. 369-374, 2010.

MOURA, C. F. H. et al. Physical and physico-chemical evaluations of acerola clones (*Malpighia emarginata Sessé e Mocino ex DC.*). **Revista Ciência Agronômica**, Fortaleza, v. 38, n. 1, p. 52-57, 2007.

MUSSER, R. D.; COUCEIRO, E. M.; ALBUQUERQUE, M. H. Effects of naphthalene acetic acid on the rooting of semi-woody acerola cuttings in a micro-sprouting system. In: BRAZILIAN FRUIT CONGRESS. 9., 1987, Campinas. **Proceedings...** Campinas: Sociedade Brasileira de Fruticultura, 1987. p. 79-83.

NACHTIGAL, J. C.; FACHINELLO, J. C.; HOFFMANN, A. Vegetative propagation by grafting. In: (Ed.). **Propagation of fruit plants**. Brasilia-DF: Embrapa Informaçao Tecnológica, 2005. p. 111-139.

NAKASONE, H. Y.; PAULL, R. E. Other american tropical fruit: acerola. In: . **Tropical fruits.** Wallingford: CABI, 1998. p. 377-389.

NASSER, M. D. et al. Vegetative propagation of different acerola cultivars by cuttings. In: SIMPÒSIO DE CIÊNCIAS DA UNESP, 6., 2010, Dracena. Proceedings **...** Dracena: UNESP, 2010.

NAUMOV, A. Tropical fruit trees of Brazil. In: CRISÓSTOMO, L. A.; NAUMOV, A. (Orgs.). **Fertilizing for high productivity and quality:** tropical fruit trees in Brazil. Fortaleza: Embrapa Agroindùstria Tropical, 2009. p. 6-12.

NESS, A. R.; POWLES, J. W. Fruit and vegetables, and cardiovascular disease: a review. **International Journal of Epidemiology**, Oxford, v. 26, n. 1, p. 1-13, 1997.

O ESTADO DE SÂO PAULO. **Capital of the acerola.** Sao Paulo: OESP, 2009. Available at: <http : //www.ibraf. org.br/news/news_item. asp?NewsID=5709>. Accessed on: May 29, 2013.

OLIVEIRA, J. R. P. et al. Botanical aspects. In: RITZINGER, R.; KOBAYASHI, A. K; OLIVEIRA, J. R.P (Eds.). **The cultivation of acerola**. Cruz das Almas: Embrapa Mandioca e Fruticultura, 2003a. p. 17-23.

OLIVEIRA, J. R. P. et al. Seedling production. In: RITZINGER, R.; KOBAYASHI, A. K; OLIVEIRA, J. R.P (Eds.). **The culture of the acerola tree**. Cruz das Almas: Embrapa Mandioca e Fruticultura, 2003b. p. 73-88.

OLIVEIRA, J. R. P.; SOARES FILHO, W. S.; CUNHA, R. B. **A cultura da acerola no Brasil**. Cruz das Almas: Embrapa Cassava and Fruit Growing, 1998. 35 p.

PAIVA, J. R. **Aceroleira clones**. Fortaleza: Embrapa Agroindùstria Tropical, 2003b. 4 p.

PAIVA, J. R. et al. **Aceroleira clones**: BRS 235 or Apodi, BRS 236 or Cereja, BRS 237 or Roxinha, BRS 238 or Frutacor. Fortaleza: Embrapa Agroindùstria Tropical, 2003. 3 p. (Technical Communication, 87).

PAIVA, J. R. et al. Genetic parameters in free-pollinated progenies of acerola (Malpighia spp). **Pesquisa Agropecuâria Brasileira**, Brasilia-DF, v. 34, n. 4, p. 629-634, 1999.

PAIVA, J. R. Seedling production. In: MANICA, I. et al. **Acerola:** production technology, post-harvest, freezing, exports, markets. Porto Alegre: Cinco Continentes, 2003a. p. 89-134.

PEDROSA, A. C. et al. Influence of the grafting process on the propagation of acerola in Pernambuco. In: CONGRESSO BRASILEIRO DE FRUTICULTURA, 13, 1994, Salvador. **Proceedings...** Salvador: Sociedade Brasileira de Fruticultura, 1994. p. 78-79.

POMPEU JÛNIOR, J. Rootstocks. In: MATTOS JÛNIOR, D. et al. (Eds.). **Citrus**. Campinas: Instituto Agronômico; Fundag, 2005. p. 63-104.

RITZINGER, R.; RITZINGER, C. H. S. P. **Acerola:** aspectos gerais da cultura. Cruz das Almas: Embrapa Cassava and Fruit Growing, 2004. 2 p. (Acerola em Foco, 9).

Available at:

<http://www.cnpmf.embrapa.br/publicacoes/produto_em_foco/acerola_09.pdf>. Accessed on: May 30, 2013.

RITZINGER, R.; RITZINGER, C. H. S. P. Acerola. **Informe Agropecuàrio**, Belo Horizonte, v. 32, n. 264, p. 17-25, 2011.

RITZINGER, R.; RITZINGER, C. H. S. P. **Propagation of the acerola tree**: grafting x cuttings. Cruz das Almas: Embrapa Cassava and Fruit Growing, 2003. 2 p. (Acerola em Foco, 4).

RITZINGER, R; SOARES FILHO, W. S. S.; OLIVEIRA, J. R. P. Varieties and improvement. In: RITZINGER, R.; KOBAYASHI, A. K.; OLIVEIRA, J. R. P. (Eds.). **The cultivation of acerola**. Cruz das Almas: Embrapa Mandioca e Fruticultura, 2003. p. 65-72.

S.A.S. INSTITUTE. **SAS procedures guide**. Version 8 (TSMO). Cary: SAS Institute, 1999.

SASSO, S. A. Z.; CITADIN, I.; DANNER, M. A. Propagation of jaboticaba by grafting and alporquia. **Revista Brasileira de Fruticultura**, Jaboticabal, v. 32, n. 2, p. 571-576, 2010.

SILVA, C. P. **Effect of naphthaleneacetic acid and indolbutyric acid on the rooting of cuttings of acerola (*Malpighia emarginata* DC.), pine (*Annona squamosa* L.) and soursop (*Annona muricata* L.), under intermittent nebulization.** 2004. 100 f. Dissertation (Master's Degree in Agronomy) - Faculty of Engineering, Universidade Estadual Paulista, Ilha Solteira, 2004.

SIMÂO, S. **Manual de fruticultura**. Sao Paulo: Agronòmica Ceres, 1971. 493 p.

SIMÂO, S. **Tratado de fruticultura**. Piracicaba: FEALQ, 1998. 760 p.

SIMPLiCIO, J. B. et al. Evaluation of the presence of embryos in seeds of two selections of acerola (*Malpighia glabra*) in the Zona da Mata of PE. In: CONGRESSO BRASILEIRO DE FRUTICULTURA, 13, 1994, Salvador. **Proceedings...** Salvador: Sociedade Brasileira de Fruticultura, 1994. p. 80.

SINGH, D. B.; ATTRI, B. L. Effect of IBA on rooting in West Indian Cherry

(*Malpighiaglabra* L.) cuttings. **Journal of Applied Horticulture**, Lucknow, v. 2, p. 134-135, 2000.

SIQUEIRA, K. M. M. et al. Comparative study of pollination in varieties of acerola trees (*Malpighia emarginata* DC, Malpighiaceae). **Revista Caatinga**, Mossoró, v. 24, n. 2, p. 18-25, 2011. Available at: <http://periodicos.ufersa.edu.br/revistas/index.php/sistema/article/view/1707/4707>. Accessed on: May 31, 2013.

SOUZA JÛNIOR, V. S. et al. Evaluation of the presence of embryos in seeds of three selections of acerola (*Malpighia glabra*) in the sertao of Pernambuco. In: CONGRESSO BRASILEIRO DE FRUTICULTURA, 13, 1994, Salvador. **Proceedings...** Salvador: Sociedade Brasileira de Fruticultura, 1994. p. 81.

SOUZA, L. D.; GOMES, J. C.; SOUZA, L. S. Soil management and conservation. In: RITZINGER, R.; KOBAYASHI, A. K; OLIVEIRA, J. R. P (Ed.). **The cultivation of acerola**. Cruz das Almas: Embrapa Mandioca e Fruticultura, 2003. p. 30-47.

STOCLET, J. C. et al. Vascular protection by dietary polyphenols. **European Journal of Pharmacology**, Amsterdam, v. 500, p. 299-313, 2004.

CAMPINAS STATE UNIVERSITY. Food Studies and Research Center - NEPA. **TACO - Table of Food Composition.** 4 ed. rev. and ampl. Campinas: NEPA, 2011. 161 p.

VIEIRA, C. P. G. et al. Propagule origin and grafting methods in the propagation of the acerola tree (*Malpighia glabra* L.). **Revista Brasileira de Fruticultura**, Cruz das Almas, v. 19, n-3, p. 347-352, 1997.

MIX
Papier aus verantwortungsvollen Quellen
Paper from responsible sources
FSC® C105338

Printed by Books on Demand GmbH, Norderstedt / Germany